Energy and Ecosystems

Lesson 1

How Do Plants Produce Food?........................ 2

Lesson 2

How Does Energy Pass Through an Ecosystem?...... 10

Orlando   Austin   New York   San Diego   Toronto   London

Visit *The Learning Site!*
www.harcourtschool.com

Lesson 1

How Do Plants Produce Food?

VOCABULARY
transpiration
photosynthesis
chlorophyll
producer
consumer

Transpiration is the loss of water through leaves.

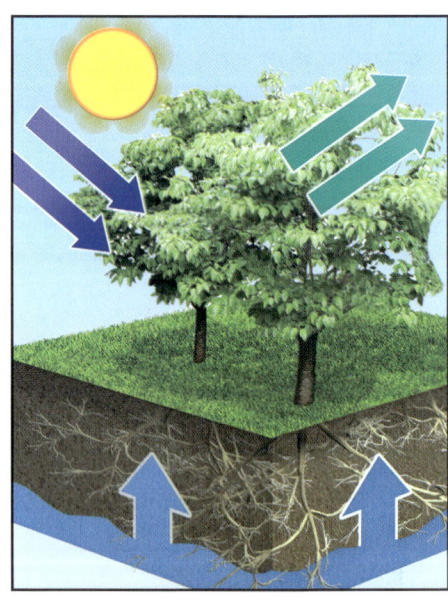

Photosynthesis is the process plants use to make food. The process uses light energy, water, and carbon dioxide.

2

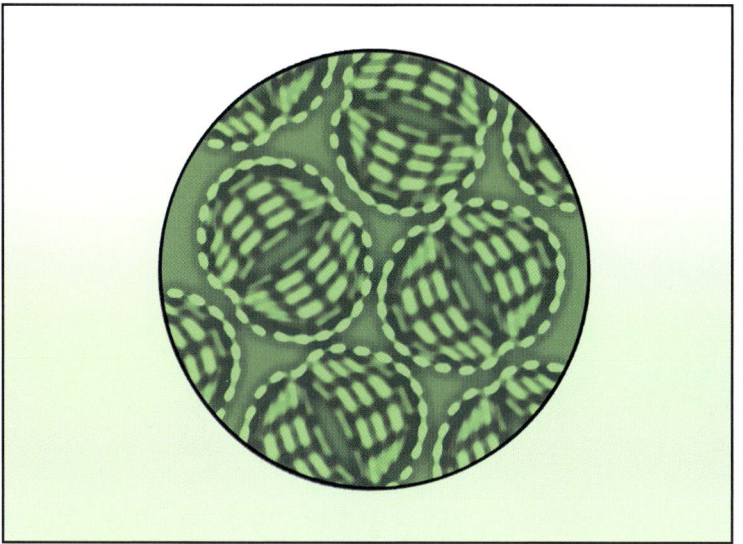

Chlorophyll is the green coloring in plants. It helps the plant use the sun's light energy.

A **producer** is a living thing that makes its own food. Plants are producers. **Consumers** are animals that eat plants, other animals, or both.

READING FOCUS SKILL
MAIN IDEA AND DETAILS

The **main idea** is what the text is mostly about. **Details** are pieces of information that tell about the **main idea**.

As you read, look for **details** about how plants make and store food.

Plant Parts

The basic parts of plants are roots, stems, leaves, and flowers. Roots anchor a plant in the ground. They also take in water and nutrients. Tubes in the roots carry water and nutrients to the stems and leaves.

Both roots and stems have tubes running through them. ▶

tubes

tubes

tubes

tubes

4

Stems support the plant and allow the leaves to reach sunlight. Stems also contain tubes that carry water and nutrients. Other tubes in stems carry food.

◀ Veins in a leaf connect to the tubes in the stem.

The main job of leaves is to make food. Most leaves have several layers of cells. The outer layer keeps the leaf from drying out. The bottom of a leaf has many small openings. These are called stomata (STOH•muh•tuh).

Stomata open during the day. This lets the leaf take in carbon dioxide to make food. At night, the stomata close. This keeps the plant from drying out. The loss of water through the stomata is called **transpiration**.

 Tell what the stomata in leaves do.

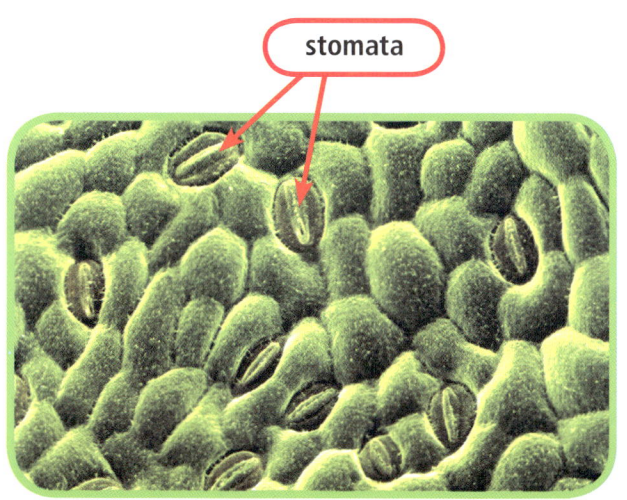

◀ When stomata are open, they let in carbon dioxide. They close at night to keep water in.

5

Photosynthesis

Plants make food they need by **photosynthesis**. The process uses water, carbon dioxide, and sunlight.

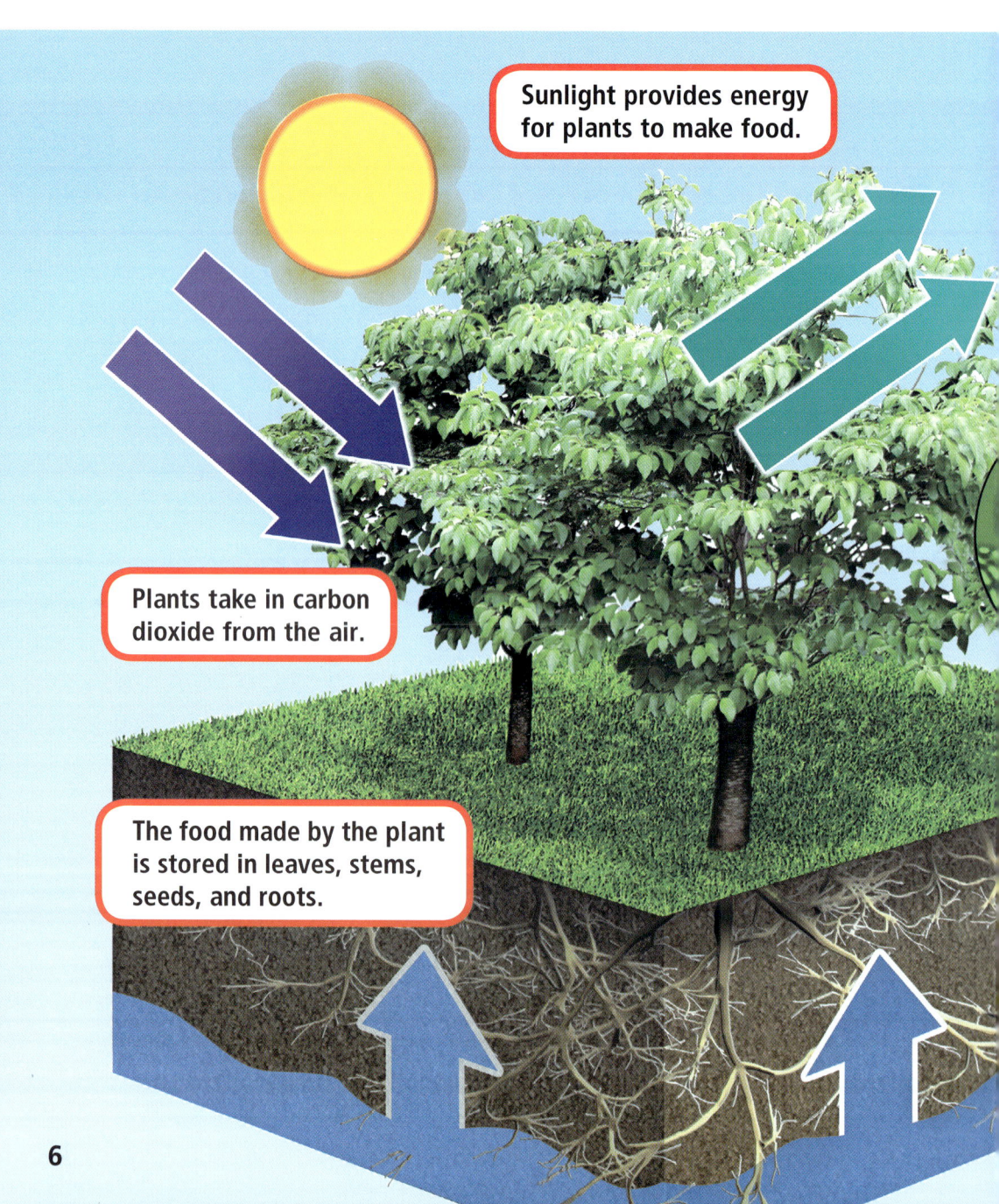

Sunlight provides energy for plants to make food.

Plants take in carbon dioxide from the air.

The food made by the plant is stored in leaves, stems, seeds, and roots.

Most plant cells contain chlorophyll. **Chlorophyll** helps a plant use light energy to make food. It also makes plants green.

Photosynthesis begins when sunlight hits the plant cells. The sun's energy is taken in. This energy causes water and carbon dioxide to make sugar. Sugar is the food that plants need to live and grow.

Oxygen is given off during photosynthesis. The oxygen is passed into the air through the stomata. The diagram shows how photosynthesis works.

Explain what a plant needs in order for photosynthesis to take place.

After making food, the leaves release oxygen.

Chlorophyll takes energy from sunlight. The plant needs this energy to make food.

Plant roots take in water, which is needed for photosynthesis.

Producers and Consumers

All living things need energy to live and grow. That energy comes from food. Plants are called **producers** because they produce, or make, food.

Animals can't make food. But they still need energy to live and grow. When animals eat plants, the animals get energy stored in the plants. The word *consume* means "to eat." So we call animals that eat plants **consumers.**

These bison get energy from eating plants. The plants are producers and the bison are consumers. Without the plants, the bison could not live.

You and every animal need plants. Even animals that eat only other animals need plants. Without plants, animals such as deer and rabbits would starve. Then animals such as wolves, which eat deer and rabbits, would have nothing to eat. They would starve, too.

The energy from sunlight moves from plants to animals that eat plants. Then it moves to animals that eat other animals. Without sunlight, nearly every living thing on Earth would die.

 Tell the difference between a producer and a consumer.

Review

 **Complete this main idea sentence.**

1. All organisms need _____ in order to live and grow.

Complete these detail statements.

2. Tubes in the roots of plants carry water and nutrients to the _____ of plants.

3. The underside of a leaf has many small openings called _____.

4. Without _____, nearly every living thing on Earth would die.

5. _____ is given off as the result of photosynthesis.

9

Lesson 2

How Does Energy Pass Through an Ecosystem?

VOCABULARY
ecosystem
herbivore
carnivore
food chain
decomposer
food web
energy pyramid

An **ecosystem** is all the living things in an area. It is also the environment in which they live. This picture shows part of a pond ecosystem.

An **herbivore** is an animal that eats plants. A **carnivore** is an animal that eats animals. This wolf is a carnivore.

A **food chain** is the transfer of food energy in an ecosystem. This caribou gets its food energy by eating moss.

A **decomposer** is a kind of consumer. It gets its food energy from dead organisms.

A **food web** shows the relationship among food chains. This is part of a prairie food web.

An **energy pyramid** shows how much food energy is passed along in a food chain.

READING FOCUS SKILL
SEQUENCE

Sequence is the order in which events take place. As you read, look for the order of events as energy passes through an ecosystem.

Energy

An <mark>ecosystem</mark> includes all the plants and animals in an area. It also includes the environment in which they live. The picture shows a tundra ecosystem. Tundra is a treeless, cold region in the Arctic.

Because plants make food, they are the producers in most ecosystems. In a tundra ecosystem, reindeer moss is a producer.

Some animals eat plants. The food energy stored in moss is passed on to the caribou. An animal that eats plants is a <mark>herbivore</mark>. Herbivores are also called first-level consumers.

Reindeer moss makes food. The food energy is stored in the moss.

The caribou gets food energy by eating reindeer moss.

12

Wolves get their food energy by eating caribou. The food energy stored in the caribou is passed to the wolf. An animal that eats other animals is called a **carnivore**. Carnivores are also called second-level consumers.

When one organism eats another, food energy is passed on. The passing of food energy in an ecosystem is called a **food chain**.

Decomposers are an important part of a food chain. **Decomposers** break down dead plants and animals. They use some of the nutrients for food. The rest of the nutrients mix with the soil. Then the roots of plants absorb them. Decomposers may be earthworms, fungi, or bacteria.

 Tell what can happen to the food energy in a first-level consumer.

The wolf gets food energy by eating caribou.

When the moss, caribou, and wolf die, decomposers break them down. Then the reindeer moss takes up any nutrients.

Food Webs

Most animals eat more than one kind of food. A hawk might eat a mouse or a small snake. A snake might eat a mouse. A mouse might eat seeds or insects. Insects might eat seeds. In this way, food chains overlap. A **food web** shows the relationship among food chains.

Carnivores eat herbivores and other carnivores. Omnivores eat both plants and animals.

Prairie Food Web

All the plants in this prairie ecosystem are producers. Herbivores include insects and bison. Carnivores include snakes and hawks. Mushrooms are decomposers.

In a healthy ecosystem, organisms need each other. They need each other to survive. If there are fewer of one kind of organism, then another organism might not eat.

 Identify some food chains that are part of a pond or prairie food web.

Pond Food Web

In this pond ecosystem the producers are water plants and algae. Herbivores include insects, tadpoles, and ducks. Carnivores include fish and some birds. The turtle is an omnivore. The water is full of decomposers, such as snails, worms, and protists.

Energy Pyramid

Not all the food energy in plants is passed on to the animals that eat them. Producers use most of their food energy for themselves. They store the other energy.

Herbivores get the energy the producers store. The herbivores use most of that energy for themselves. They store the other energy.

An **energy pyramid** shows how much food energy is passed through a food chain. The base of the pyramid is large. It shows where most of the food energy is. Each layer of the pyramid becomes smaller. Less and less energy is passed along.

If any part of the food chain changes, the entire food chain is affected. Suppose a lack of rain kills most of the plants in an area. Some of the next consumers will starve. Then many other consumers may starve, too.

 Explain what may happen if one level of a food chain is changed.

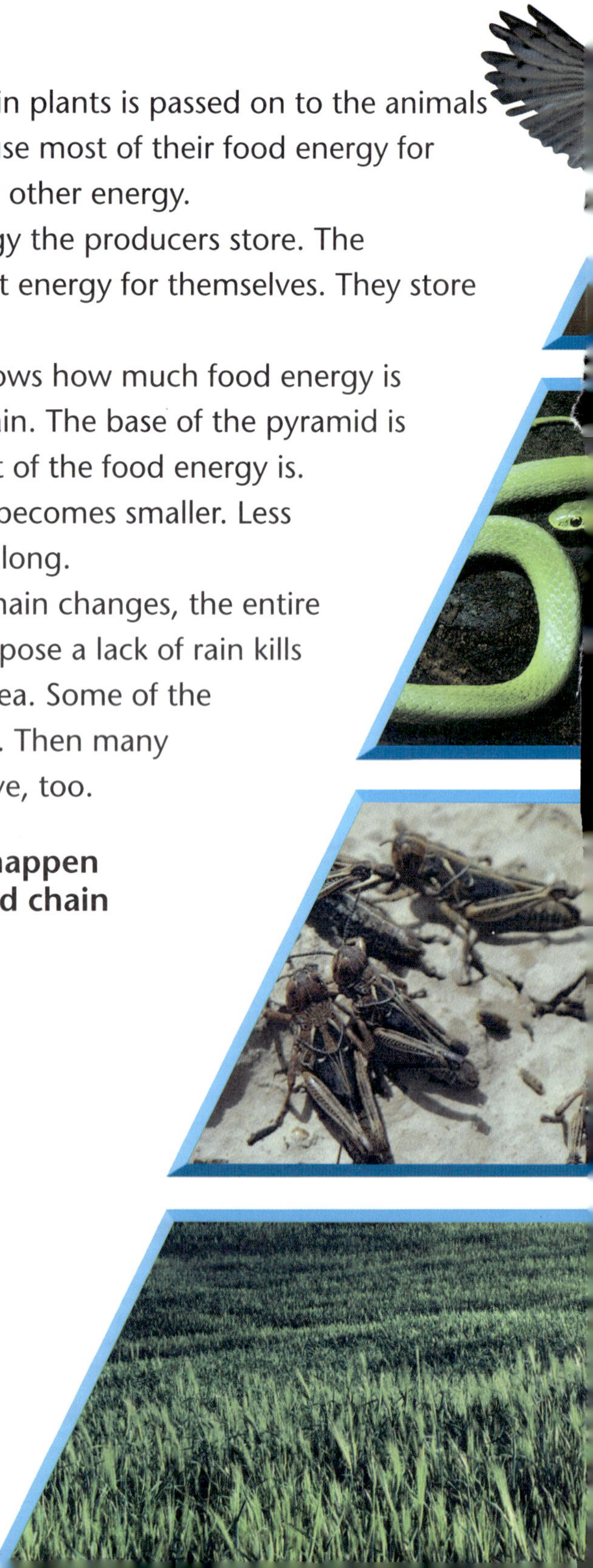

Owl

The owl is a third-level consumer. It takes a lot of grass, locusts, and snakes to feed the owl.

Snakes

The snakes are second-level consumers. They pass on some of the energy they get from the locusts.

Locusts

The locusts are first-level consumers. They pass on some of the energy they get from the grasses.

Grasses

The grasses are producers. They pass on some of the energy they produce.

Natural Cycles

Most ecosystems depend on the water cycle. The water cycle supplies plants with water for photosynthesis.

Nitrogen also has a cycle. Nitrogen is important for all living organisms. Nitrogen is a gas that makes up most of Earth's atmosphere. But nitrogen gas cannot be used by plants. It must be changed so plants can take it up through their roots.

Lightning changes some nitrogen into another form. Plant roots can use this form.

Plants use nitrogen to grow.

Lightning changes nitrogen.

Animals eat plants that contain nitrogen.

Animal wastes put nitrogen back into the soil.

Some plants have bumps on their roots. These contain bacteria. The bacteria change nitrogen into a form the plant can use.

When a plant dies, nitrogen is put back into the soil. Animal wastes also contain nitrogen. Decomposers change these wastes into nitrogen that plants can use.

Bacteria on some plant roots change nitrogen into a form plants can use.

 Tell two ways nitrogen gets into soil.

Review

 **Complete these sequence statements.**

1. _____ are at the first level of a food chain.

2. Food energy passes from producers to _____ consumers.

3. Food energy passes from first-level consumers to _____ consumers.

4. As energy moves up the energy pyramid, _____ and _____ energy is passed along.

GLOSSARY

carnivore (KAHR•nuh•vawr) An animal that eats other animals. Also called a second-level consumer.

chlorophyll (KLAWR•uh•fihl) A green coloring that helps a plant use the sun's light energy.

consumer (kuhn•SOOM•er) An animal that eats plants, other animals, or both.

decomposer (dee•kuhm•POHZ•er) A consumer that gets food energy from dead plants and animals.

ecosystem (EE•koh•sis•tuhm) Plants and animals and the environment they live in.

energy pyramid (EN•er•jee PIR•uh•mid) A diagram that shows how much food energy is passed along in a food chain.

food chain (FOOD CHAYN) The transfer of food energy in an ecosystem.

food web (FOOD WEB) A diagram that shows the relationships among food chains.

herbivore (HER•buh•vawr) An animal that eats only plants.

photosynthesis (foht•oh•SIHN•thuh•sis) The process plants use to make food. They use water from the soil, carbon dioxide from the air, and energy from sunlight.

producer (pruh•DOOS•er) A living thing, such as a plant, that makes its own food.

transpiration (tran•spuh•RAY•shuhn) The loss of water from a leaf.